Synthetic & Renewable Fuels 2019

Topical Conference at the 2019 AIChE Annual Meeting

Orlando, Florida, USA
10-15 November 2019

ISBN: 978-1-7138-0565-6

Printed from e-media with permission by:

Curran Associates, Inc.
57 Morehouse Lane
Red Hook, NY 12571

Some format issues inherent in the e-media version may also appear in this print version.

Copyright© (2019) by AIChE
All rights reserved.

Printed with permission by Curran Associates, Inc. (2020)

For permission requests, please contact AIChE
at the address below.

AIChE
120 Wall Street, FL 23
New York, NY 10005-4020

Phone: (800) 242-4363
Fax: (203) 775-5177

www.aiche.org

Additional copies of this publication are available from:

Curran Associates, Inc.
57 Morehouse Lane
Red Hook, NY 12571 USA
Phone: 845-758-0400
Fax: 845-758-2633
Email: curran@proceedings.com
Web: www.proceedings.com

TABLE OF CONTENTS

Shell Gas-to-Liquids

Joseph B. Powell[1] and Leendert Bezemer[2]

Abstract:

An overview of Shell Gas-to-Liquids technology is presented, including R&D activities dating from 1973, and scale-up through an initial plant in Bintulu, to full scale deployment as project Pearl (Qatar 2011).

GTL Overview:

As the largest of commercial Gas-to-Liquids or "GTL" technology [1], Shell's Middle Distillates Synthesis or "SMDS" [2-5] converts natural gas to higher-value liquid fuels and chemicals via a multi-step process (Figure 1). Methane is first converted to synthesis gas via partial oxidation [6] in the "Shell Gasification Process [7] or SGP":

$$CH_4 + \frac{1}{2} O_2 \rightarrow CO + 2H_2$$

Syngas is then reacted over a proprietary catalyst in multi-tubular fixed-bed reactors to form middle distillates using Fischer-Tropsch chemistry [8] with high probability for chain growth, in the Heavy Paraffins Synthesis (HPS) step :

$$nCO + mH_2 \rightarrow \text{paraffins} + zH_2O$$

This provides for relatively low yield of products in the gas and gasoline range. The heavy end of the Schulz-Flory product distribution is then hydrocracked (Heavy Paraffin Conversion or HPC) to optimize product yield across fuels, solvents, and lube-oil applications [9].

Figure 1: Shell GTL

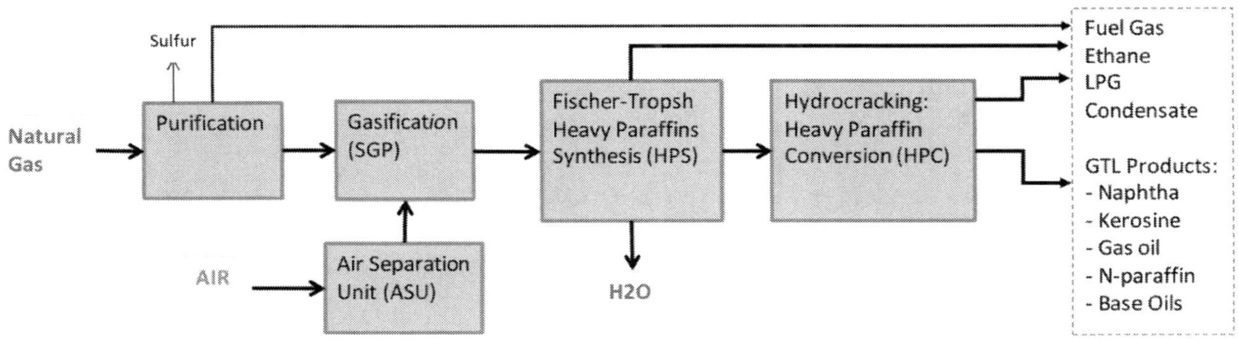

[1] Chief Scientist, Shell International Exploration and Production (US) Inc.
[2] Shell Global Solutions International

History and Scale:

Technology development was scaled via a pilot plant at Shell's Amsterdam lab in the 1970's. A 14,700 barrel per day (bpd) small commercial plant in Bintulu was subsequently constructed to produce GTL fuel, chemical feedstock and a range of specialties including waxes and drilling fluids, with first product shipments in 1993. The world's largest GTL plant was then constructed as project Pearl in Qatar, producing 120,000 barrels of oil equivalent per day (boe/d) of natural gas liquids and ethane, and 140,000 bbl/d of liquid hydrocarbon GTL products. First cargo was loaded on 13th June 2011.

The Pearl plant contains the world's largest water treatment facility for recovering, processing and re-using process water with zero liquid discharge capability, and the world's largest air separation unit (ASU) with 28,800 t/d capacity. Produced water from Fisher-Tropsch synthesis renders the plant self-sufficient in water. Paraffin products have been found to exert minimal adverse ecotoxicological and health effects as a sustainability benefit relative to conventional petroleum-derived analogs.

References:

[1] D. A. Wood, C. Nwaoha and B. F. Towler, "Gas-to-liquids (GTL): A review of an industry offering several routes for monetizing natural gas," *Journal of Natural Gas Science and Engineering,* vol. 9, pp. 196-208, 2012.

[2] J. Eilers, S.A. Posthuma, S.T. Sie, The shell middle distillate synthesis process (SMDS), Catal Lett. 7 (1991) 253–269. doi:10.1007/BF00764507.

[3] P. Tijm, "SHELL MIDDLE DISTILLATE SYNTHESIS: THE PROCESS, THE PLANT, THE PRODUCTS," *ACS Fuel Preprint,* vol. 39, p. 1146, 1994.

[4] J. Ansorge, "Shell Middle Distillate Synthesis: Fischer-Tropsch Catalysis in Natural Gas Conversion to High Quality Products," *Preprint Paper: Am. Chem. Society Div. Fuel Chemistry,* vol. 43, no. 2, pp. 654-663, 1997.

[5] M. J. van den Burgt, C. J. van Leeuwen, J. J. del Amico and S. T. Sie, "The Shell Middle Distillate Process," in *Methane Conversion*, D. Bibby, C. Chang, R. F. Howe and S. Yurchak, Eds., Amsterdam, Elsevier, 1988, pp. 473-482.

[6] S.T. Sie, M.M.G. Senden, H.M.H. Van Wechem, Conversion of natural gas to transportation fuels via the shell middle distillate synthesis process (SMDS), Catalysis Today. 8 (1991) 371–394.

[7] C. Higman and M. van der Burgt, Gasification, Elsevier, 2008.

[8] H. H. Storch, N. Gulombic and R. B. Anderson, The Fischer-Tropsch and Related Syntheses, NY: John Wiley and Sons, 1961.

[9] [7] S. Gullapalli and A. Falender, "GTL innovation produces clean base oils from natural gas", Gas Processing and LNG, http://gasprocessingnews.com/features/201410/gtl-innovation-produces-clean-base-oils-from-natural-gas.aspx (accessed 27 Sept 2019).

[10] R. Overtoom, N. Fabricius and W. Leenhouts, "Shell GTL, from Bench Scale to World Scale," in Proceedings of the 1st Annual Gas Processing Symposium, Elsevier, 2009, pp. 378-386.

581754 Methanol – a Practical Transportation Fuel for Today and Tomorrow

The presentation will discuss developing applications of methanol as clean transportation fuel produced from natural gas or coal, and increasingly from renewable resources including captured CO2 and renewable electricity. Out of ~80 million metric tons of methanol sold globally in 2018, energy and fuel uses represent 40% of total demand. Use of methanol as vehicle fuel is rapidly expanding in China and being tested in the EU and other parts of the world. Methanol applications in high-octane blending with gasoline; as rapidly expanding clean burning marine fuel; and in range expanding applications for fuel cell cars will be discussed.

Copyright © American Institute of Chemical Engineers. All rights reserved.

581784 Green Technology Solutions for Synthetic Fuels

The paper will present a study for alkaline water electrolysis (AWE) based ammonia, methanol and SNG production units using only a minimum number of equipment in order to reduce the investment cost.

Currently three trends can be observed leading to expectation of increasing demand for small-scale green chemical plants in the future.

First, focusing primarily on ammonia, safety and security considerations have led to transport restrictions affecting the reliability of supply leading to increased transportation costs for the supply of green chemicals. This has become an issue for sites that currently get their ammonia supply by rail or ship.

Second, there is an increasing share of electricity produced from renewable sources which are available in inconsistent quantities (e.g. wind and solar) sometimes exceeding demand. This surplus power can be used to produce hydrogen via water electrolysis. In case of ammonia, nitrogen can be supplied by an air separation unit. Such production of ammonia is free of carbon dioxide emissions, and can be considered "green ammonia". In case of methanol and SNG CO_2 waste streams of chemical plant complexes or conventional power plants can be utilized. Such production of methanol and SNG reduces the carbon footprint of adjacent chemical complexes and power plants. These "green chemicals" can be used either as feedstock for other chemical processes or as a chemical energy storage in times of high energy production from renewable sources and low network load.

Third, new regulations implementing CO_2 quotas for natural gas usage make natural gas based productions in some regions less attractive.

Copyright © American Institute of Chemical Engineers. All rights reserved.

581758 Innovation for Direct Use of Ammonia in the Energy Market

SIP "Energy Carriers" in Japan from 2014 to 2018 developed technologies to use ammonia directly in power generation, industrial and maritime markets.

Key achievements are developments of Gas Turbines from small size to ACCGT, SOFC, mix combustion in coal fired power units, industrial furnaces and maritime diesel engines. Technologies to control NOx emissions under regulations are well developed.

Our feasibility studies indicate high potentiality of ammonia for carbon-free fuel in the energy market.

After SIP "Energy Carriers" was finished, further technology development, demonstration and deployment would be continuously supported by The Green Ammonia Consortium established April 2019. The consortium is targeting to start commercial use of carbon-free ammonia in power generation in the middle of 2020s.

Copyright © American Institute of Chemical Engineers. All rights reserved.

581763 RE-Fueling the Future: Hydrogen As Critical Intermediate in Renewable Energy Conversion into Fuels

Synthetic renewable fuels (RE-fuels) can be produced through reduction of captured CO_2 or N_2 utilizing wind, solar and/or other forms of renewable energy. All the technologies needed for such renewable fuels production, namely, renewable electricity generation, water electrolysis, carbon capture, and liquid fuel synthesis already exist and are practiced on a commercial scale. Further R&D efforts should link these technologies together and ensure cost reduction and market acceptance. Cost analysis indicates that the cost of hydrogen production constitute the major fraction of the overall cost of renewable fuel. Lowering cost of hydrogen as part of H2@Scale Initiative is critical for making RE-fuels cost competitive with fossil alternatives.

Copyright © American Institute of Chemical Engineers. All rights reserved.

Future Role of Hydrogen in Dense Energy Carriers and Energy Storage

Joseph B. Powell Ph.D. (Shell Chief Scientist, Chemical Engineering)

Abstract

The multiple roles for hydrogen as a dense energy carrier and for renewable energy storage will be presented and contrasted with other competing options for future mobility, power, and industrial energy. Current progress in roll out of hydrogen as key energy vector will also be reviewed.

Technical Summary

A hydrogen economy has been proposed for multiple decades [1, 2]. Hydrogen is not a primary energy source, but as an energy vector (like electricity, or the proposed "methanol economy"), it has a number of attractive features. Once generated, hydrogen can be used in a fuel cell at high efficiency, avoiding the Carnot inefficiency of fuels combustion. Its fuel cell or primary combustion product is water, and hence it can be used in low- or zero-emission vehicles and power applications [3]. Increased concern over local air quality (criteria pollutants) and climate change risk (CO_2 and other greenhouse gas emissions) has driven increased interest in hydrogen as an energy vector [3-18].

Renewable energy power generation costs from wind and solar have decreased dramatically over the last decade [19]. Wind and solar and are contributing 10% of power generation as of 2018 [20], but intermittency leads to oversupply and market irregularities such as the "duck curve" [21]. Battery electric vehicles (BEV) have seen growing deployment as a means to reduce carbon footprints and emissions for future low carbon power generation [22].

While instantaneous power generation costs from wind and solar are inexpensive, storage of that energy in the form of batteries is not [23]. Small amounts of intermittent power production can be managed by grid control, but larger percentages will require storage solutions. Many urban markets are not in regions of high intensity solar or wind resource. Hydrogen as a molecular energy carrier can allow both transport and storage of renewable (green) energy to market (Figure 1), to address larger penetration of renewables into the energy grid. In addition, clean or "blue" hydrogen can be produced from natural gas via steam methane reforming, partial oxidation, or pyrolysis combined with sequestration of carbon [24]. Hydrogen is thus a common energy vector for both renewable energy, and clean energy derived from fossil resources 25]. The hydrogen can be used in mobile of stationary fuel cells to provide electric power for vehicles or commercial and residential applications, or combusted to provide low- or zero-emission industrial or even residential energy (as proposed for UK).

Infrastructure for generation, storage, and transport of hydrogen is required for its use as an energy vector, unlike renewable power which can be placed into existing grids at low percent usage. Costs are reduced via increasing total scale of deployment, by sharing applications across industry and personal use sectors. One of most successful deployments to date is in fuel cell lift trucks (forklifts), where more than 31,000 have been deployed [26]. Fuel cell fleets are more economic than battery electric for this application, due to rapid refueling and hence increased stream factor for utilization. Hydrogen fuel cells are also advantaged for decarbonization of heavy duty transport and shipping, as well as future aviation

[4-6, 12-14]. Creation of shared infrastructure and demand for larger scale production in order to drive down costs is the current mandate for the US DOE's H2@Scale initiative, and for other global energy strategies [9-18].

Figure 1: Hydrogen as an Energy Vector

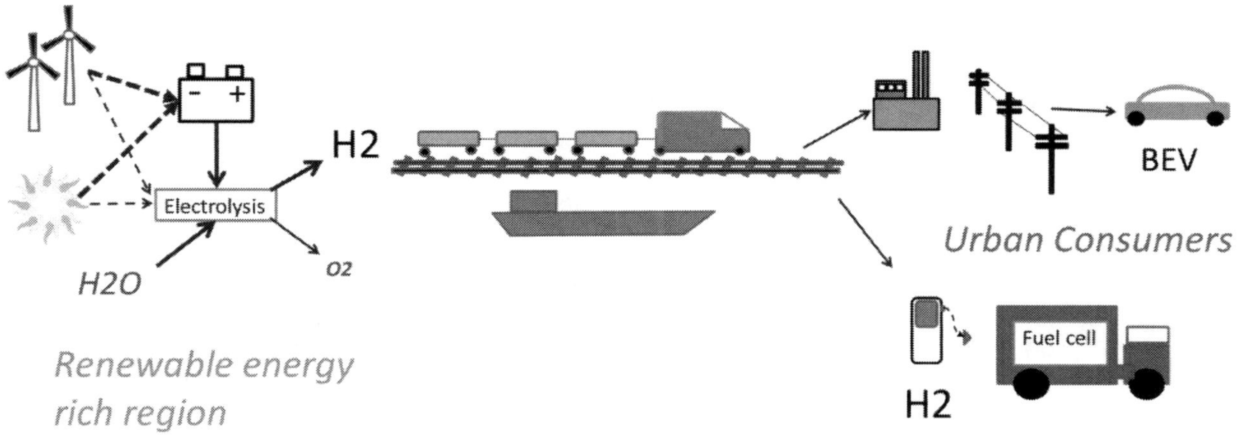

Hydrogen if used as an energy vector to move renewable (green) or clean (blue) energy to market, also provides storage and synergy with use of fuel cells for heavy and light duty transit, as well as future aviation, while also enabling decarbonization and local emission reduction for industry.

References

1. Bockris, J. O. M. The Hydrogen Economy: Its History. Int. J. Hydrogen Energy 2013, 38 (6), 2579–2588. https://doi.org/10.1016/j.ijhydene.2012.12.026.

2. G.W. Crabtree, M.S. Dresselhaus, M.V. Buchanan, The Hydrogen Economy, Physics Today. 57 (2004) 39–44. doi:10.1063/1.1878333

3. Elgowainy, A., J. Han, J. Ward, F. Joseck, D. Gohlke, A. Lindauer, T. Ramsden, M. Biddy, M. Alexander, S. Barnhart, I. Sutherland, L. Verduzco, T.J. Wallington, Current and Future United States Light-Duty Vehicle Pathways: Cradle-to-Grave Lifecycle Greenhouse Gas Emissions and Economic Assessment, Environ. Sci. Technol. 52 (2018) 2392–2399. doi:10.1021/acs.est.7b06006

4. Shell, ***Sky Scenario*** (2018). https://www.shell.com/energy-and-innovation/the-energy-future/scenarios/shell-scenario-sky.html

5. Shell (2017). *Shell Hydrogen Study: Energy of the Future?* *www.shell.com/hydrogen*

6. Staffell, I., D. Scamman, A. Velazquez Abad, P. Balcombe, P.E. Dodds, P. Ekins, N. Shah, K.R. Ward, The role of hydrogen and fuel cells in the global energy system, Energy Environ. Sci. 12 (2019) 463–491. doi:10.1039/C8EE01157E.

7. G. Bean, The role of hydrogen in de-carbonization of the energy industry, UH Energy White Paper Series: No. 01.2019. http://www.uh.edu/uh-energy/research/white-papers/white-papers-files/white-paper-series-1-2019.pdf

8. Muradov, N. Z.; Veziroğlu, T. N. "Green" Path from Fossil-Based to Hydrogen Economy: An Overview of Carbon-Neutral Technologies. Int. J. Hydrogen Energy 2008, 33 (23), 6804–6839. https://doi.org/10.1016/j.ijhydene.2008.08.054.

9. The Hydrogen Council. http://hydrogencouncil.com/

10. IEA, Technology Roadmap: Fuel Cells and Hydrogen (2015). https://www.iea.org/publications/freepublications/publication/TechnologyRoadmapHydrogenandFuelCells.pdf

11. J. M. Ogden, Prospects for Hydrogen in the Future Energy System, UCD-ITS-RR-18-07, March 2018. https://steps.ucdavis.edu/wp-content/uploads/2018/02/OGDEN_ProspectforHydrogen2018_UCD-ITS-RR-18-07.pdf

12. The Fuel Cell Industry Review, E4tech (2018), http://www.fuelcellindustryreview.com.

13. Hydrogen Roadmap Europe (2019): https://www.fch.europa.eu/news/hydrogen-roadmap-europe-sustainable-pathway-european-energy-transition

14. IEA. The Future of Hydrogen: Seizing Today's Opportunity (2019). Report prepared for G20. https://www.g20karuizawa.go.jp/assets/pdf/The%20future%20of%20Hydrogen.pdf

15. Committee on Climate Change (UK, 2018): *Hydrogen in a low-carbon economy.* https://www.theccc.org.uk/wp-content/uploads/2018/11/Hydrogen-in-a-low-carbon-economy.pdf

16. F2HCFuturegen. http://www.h2fcsupergen.com

17. IEA (2017: Global Trends and Outlook for Hydrogen. http://ieahydrogen.org/pdfs/Global-Outlook-and-Trends-for-Hydrogen_Dec2017_WEB.aspx

18. EERE, "H2@Scale," U.S. Department of Energy Fuel Cell Technologies Office, 2019. [Online]. Available: https://www.energy.gov/eere/fuelcells/h2scale. [Accessed 15 March 2019].H2@Scale (2019).

19. IRENA (2018). Renewable Power Generation Costs. https://www.irena.org/-/media/Files/IRENA/Agency/Publication/2018/Jan/IRENA_2017_Power_Costs_2018.pdf

20. EIA (2019): EIA forecasts renewables will be fastest growing source of electricity generation https://www.eia.gov/todayinenergy/detail.php?id=38053.

21. USDrive (2017). Roadmap. https://www.energy.gov/sites/prod/files/2017/11/f39/EETT%20Roadmap%2010-27-17.pdf

22. EERE (Office of Energy Efficiency & Renewable Energy): Confronting the Duck Curve: How to Address Over-Generation of Solar Energy, OCTOBER 12, 2017. https://www.energy.gov/eere/articles/confronting-duck-curve-how-address-over-generation-solar-energy

23. W. Cole and A. W. Frasier, Cost Projections for Utility-Scale Battery Storage. NREL 2019. https://www.nrel.gov/docs/fy19osti/73222.pdf

24. B. Parkinson, P. Balcombe, J. Speirs, A. D. Hawkes and K. Hellgardt, "Levelized cost of CO2 mitigation from hydrogen production routes," Energy and Environmental Science, vol. 12, pp. 19-40, 2019.

25. D. Parra, L. Valverde, F.J. Pino, M.K. Patel, A review on the role, cost and value of hydrogen energy systems for deep decarbonisation, Renewable and Sustainable Energy Reviews. 101 (2019) 279–294. doi:10.1016/j.rser.2018.11.010.

26. J. B. Powell et al., Annual Report of Hydrogen and Fuel Cell Technical Advisory Committee 2018), in review (2019). https://www.hydrogen.energy.gov/advisory_htac.html

581765 H2@Scale Evaluation of Market Opportunities for Water Splitting with Low Cost Clean Energy

Technical and economic evaluations of market opportunities for hydrogen production by polymer electrolyte membrane (PEM) and high temperature solid oxide electrolysis cell (SOEC) plants have shown hydrogen can be produced for under 2.00 USD per kilogram hydrogen in centralized or distributed plants. This is attributed to three factors; 1) the PEM or SOEC electrolysis plants are powered by low-cost renewable energy during periods of excess generation or by existing nuclear power plants that are mainly dedicated to hydrogen production, 2) the production of cutting-edge PEM and SOEC electrolysis stacks and supporting balance of plant and hydrogen delivery technology reaches a high volume of production, providing 100-200 MW of electrolysis capacity each year, and 3) demand for clean hydrogen increases proportionally to the growth in production. A cross-cutting DOE program analysis team dedicated to the H2@Scale Concept has shown there is an increasing market opportunity to produce and use a high volume of low-cost hydrogen in the transportation and industry sectors. Steel manufacturing, ammonia-based fertilizers, petroleum refining, and synthetic fuels produced by CO_2 with H_2 as feedstock could benefit from competitively priced, clean hydrogen. The technical and economic assessments indicate that the hydrogen produced by PEM and SOEC using clean energy sources can be competitive with incumbent steam-methane reforming.

Copyright © American Institute of Chemical Engineers. All rights reserved.

581766 Advancements in PEM Electrolyzers and Uses in the Renewable Energy Economy

PEM-based electrolyzers are at the leading edge of a global energy system transformation and are proving to be an enabling technology for renewable energy management; peak shaving, storage, pollution control, and greenhouse gas reduction. Correspondingly, this vision of a renewable energy economy, in parallel with generator-to-grid pathways, incorporates abundant renewable energy which is transformed first to hydrogen, then onward to chemical potential energy storage. In this future, the myriad chemicals, fertilizers, and fuels (both liquid as well as gaseous) that make modern life convenient, healthful, and productive, are manufactured while achieving a low–even zero carbon footprint. The presentation addresses the recent advancements in Giner ELX's MW-scale PEM-based electrolyzer technology, projects and commercial opportunities for direct electrolyzer integration with renewable power sources, and cost reduction expectations with increasing scale-up of the technology.

Copyright © American Institute of Chemical Engineers. All rights reserved.

581768 Renewable Hydrogen & Methane Production – Integration, Optimization and Cost Improvements

The move to renewable electricity sources is happening on a global scale to decarbonize the electricity sector. In a growing number of states and regions of the world, inherently variable renewable electricity sources continue to grow, but can be curtailed significantly when production outpaces demand. While MW-scale batteries do a good job of storing some of this electricity for hours to days and have a high roundtrip efficiency, storing this energy in molecular form (e.g., hydrogen or methane) provides a long-duration energy storage solution with essentially endless capacity. And in the case of renewable methane, the existing infrastructure can store terawatt-hours of energy...today. If the hydrogen is produced using renewable electricity and, in the case of methane, recycled carbon dioxide (CO2), renewable energy storage solutions become a sustainable and near-term solution that enable future expansion of electricity sources generated by solar and wind. Working closely with the Department of Energy's Fuel Cell Technologies and Bioenergy Technology offices, industry partners and other stakeholders over the past 15 years, the National Renewable Energy Laboratory (NREL) has conducted research and development in improving the cost of integrating renewable electricity sources with hydrogen-producing low-temperature electrolyzers. Today, NREL operates an electrolyzer test facility capable of power electrolyzer stacks up to 1MW with compression and storage systems up to 900 bar. In addition, the integrated test facility is host to the first-of-its kind biomethanation system that use organisms to produce methane from recycled CO2 and the renewable hydrogen produced onsite at NREL. This presentation will summarize our work in the areas of integration with renewable electricity sources, power conversion, stack performance, balance of plant optimization and performance data from the biomethanation system.

Copyright © American Institute of Chemical Engineers. All rights reserved.

581775 Department of Energy's Carbon Capture and Utilization R&D Portfolio and Applications to Renewable and Synthetic Fuels

The Department of Energy's Carbon Capture and Carbon Utilization program, administered by the Office of Fossil Energy and implemented at the National Energy Technology Laboratory, is conducting research and development activities on advanced carbon capture and utilization technologies to advance safe, cost effective, capture and storage and/or use of CO2. This talk will provide an introduction to the portfolio's capture projects which aim to provide step-change reductions in both cost and energy penalty as compared to currently available technologies for the potential carbon dioxide feedstock for renewable and synthetic fuels production. In addition, the carbon utilization program will review projects producing end-use renewable fuels and precursors such as CO. These projects are implementing novel conversion pathways and approaches via diversifying energy sources, developing new catalysts, integrating system process designs, and more.

Copyright © American Institute of Chemical Engineers. All rights reserved.

581776 Capturing CO_2 from the Atmosphere

Carbon Capture technology has focused on large point sources which account for 60% of total CO2 emissions, but most CO2 emissions are not amenable to capture at the source. Direct Air Capture (DAC) removes CO2 from the atmosphere and is a negative-emission technology. Carbon Engineering has developed a DAC process that will capture 1 million metric-tons of CO2 per year in a commercial scale plant. This presentation will discuss the unique technical challenges of DAC, the technologies developed at CE for initial commercialization, and the utilization pathways for the CO2 produced.

Copyright © American Institute of Chemical Engineers. All rights reserved.

581777 Sustainable and Renewable Carbon and Nitrogen Cycles for Fuel and Crop Production

Sustainable and Renewable Carbon and Nitrogen Cycles for Fuel and Crop Production

Daniel G. Nocera

Patterson Rockwood Professor of Energy

Harvard University

Cambridge, MA 02138 USA

Hybrid biological | inorganic (HBI) constructs have been created to use sunlight, air and water (as the only starting materials) to accomplish carbon and nitrogen fixation, thus providing a path to a sustainable nitrogen and carbon cycle for distributed and renewable fuels and crop production.

The carbon and nitrogen fixation cycles begin with the artificial leaf, which was invented to accomplish the solar process of natural photosynthesis – the splitting of water to hydrogen and oxygen using sunlight – under ambient conditions. To create the artificial leaf, an oxygen evolving complex of Photosystem II was mimicked, the most important property of which was the self-healing nature of the catalyst. Self-healing catalysts of the artificial leaf permit water splitting to be accomplished under benign conditions and thus the system may be easily interfaced with bioorganisms. To this end, using the tools of synthetic biology, a bio-engineered bacterium converts carbon dioxide from air, along with the hydrogen produced from the catalysts of the artificial leaf, into biomass and liquid fuels, thus closing an entire artificial photosynthetic cycle. The HBI, called the Bionic Leaf, operates at unprecedented solar-to-biomass (10.7%) and solar-to-liquid fuels (6.2%) efficiencies, greatly exceeding the 1% efficiency of natural photosynthesis.

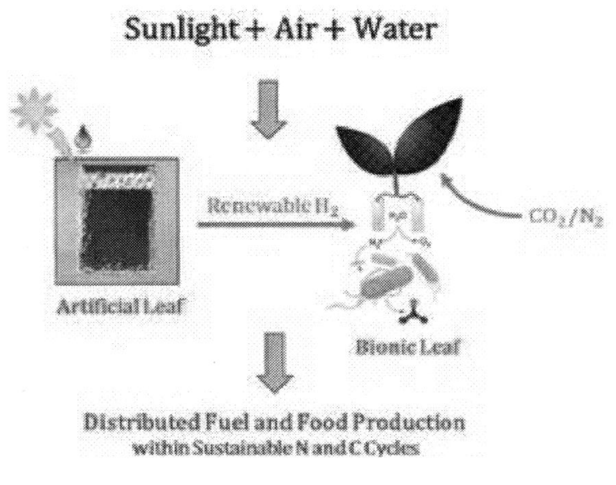

Extending this approach, we have discovered a renewable and distributed synthesis of ammonia (and fertilizer) at ambient conditions by coupling solar-based water splitting to a nitrogen fixing bioorganism in a single reactor. Nitrogen is fixed to ammonia by using the hydrogen produced from water splitting to power a nitrogenase installed in a bioorganism. Nitrogen reduction reaction proceeds at a turnover number of 10^{10} per cell and operates without the need for a carbon feedstock (other than the CO_2 provided from air). This nitrogen fixing HBI can be powered by distributed renewable electricity, enabling sustainable crop production with a *large and negative carbon budget*.

The science that will be presented will show that using only sunlight, air and water, distributed and renewable systems may be designed to produce fuel (carbon neutral) and food (carbon negative) within sustainable cycles for the biogenic elements.

Copyright © American Institute of Chemical Engineers. All rights reserved.

581778 Carbon Capture and Conversion: The Methanol Economy[R]

Methanol is preferable hydrogen gas for energy storage and transportation. It is a drop-in fuel for IC engines, and for direct oxidation fuel cells. Dimethyl ether derived from methanol is a high cetane diesel substitute and also could replace liquefied natural gas (LNG) and liquefied petroleum gas (LPG). Methanol is a convenient feedstock to produce, hydrogen, ethylene and propylene. Chemical recycling of excess carbon dioxide formed from human activities, natural and industrial sources, or even from the air can be converted to methanol via capture followed by reductive conversion with hydrogen. Any available energy source (preferably alternative energies such as solar, wind, geothermal, atomic, etc.) can provide the needed energy for generating hydrogen. The production of renewable methanol is already being demonstrated in the George Olah Methanol plant in Iceland by Carbon Recycling International employing geothermal energy. The Methanol Economy concept is expected to solve the energy and material problems of the world in the long run and at the same time address the issue of global warming due to increased CO_2 emissions by excessive fossil fuel use by efficient "*Carbon Neutral Energy Cycles.*" Our recent efforts on carbon capture and conversion will be discussed.

Copyright © American Institute of Chemical Engineers. All rights reserved.

581781 Technical and Economic Feasibility of Electron-Driven CO_2 Reduction

Renewable electricity can be leveraged to produce fuels and chemicals from CO_2, offering sustainable routes to reduce the carbon intensity of our energy and products-driven economy. A number of approaches have been developed for the electron-driven reduction of CO_2 to products, including both direct and indirect (via an energy carrier such as H_2) pathways and spanning from electrochemical to biological to thermocatalytic conversion. This presentation will discuss the technical and economic feasibility of these pathways for utilizing electricity to reduce CO_2 into C_1-C_3 products. The integration of these CO_2 utilization technologies with emerging biorefineries will also be discussed.

Copyright © American Institute of Chemical Engineers. All rights reserved.

AUTHOR INDEX